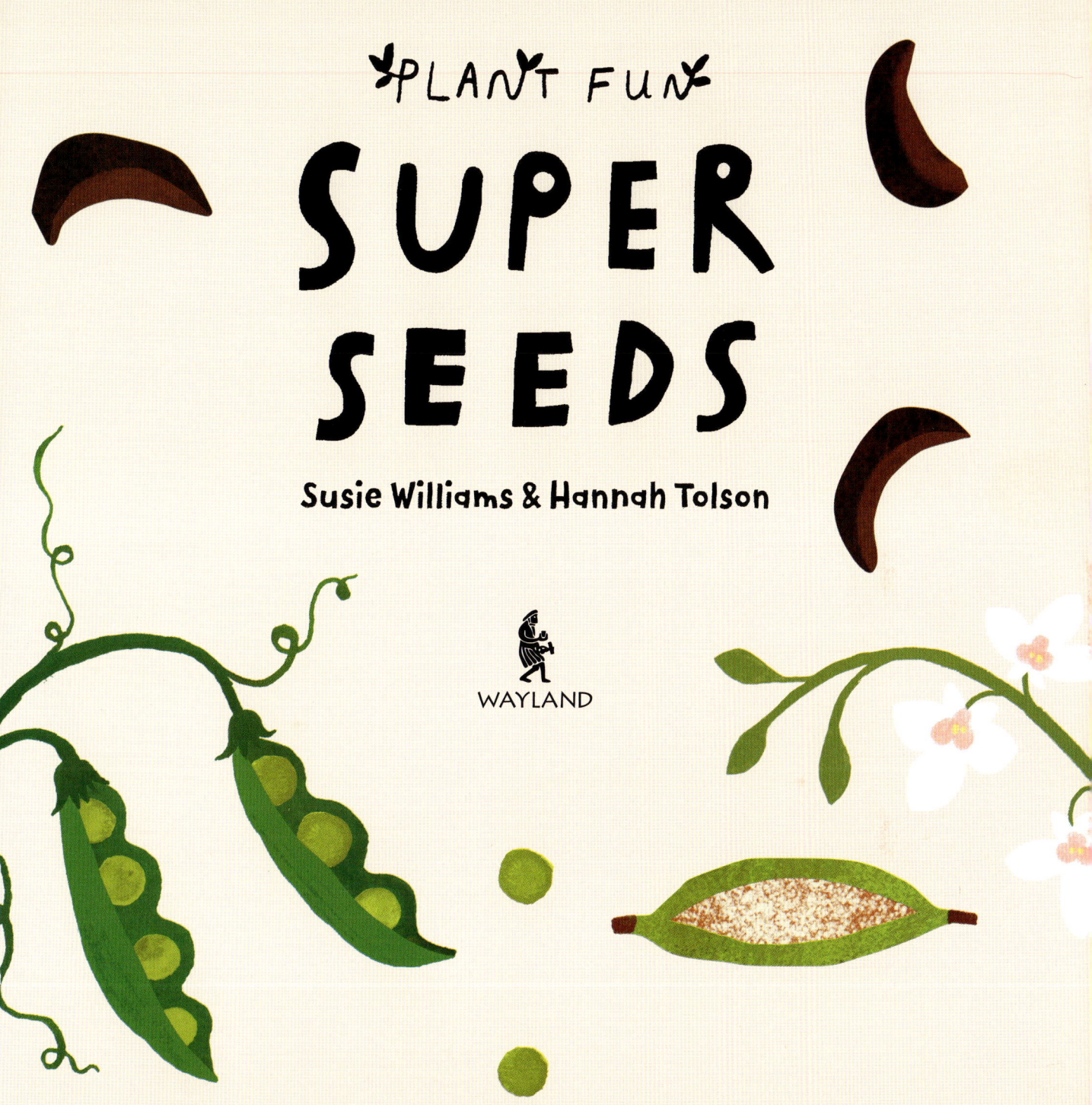

PLANT FUN

SUPER SEEDS

Susie Williams & Hannah Tolson

WAYLAND

First published in Great Britain in 2025
by Hodder & Stoughton

Published in association with the Royal Horticultural Society, London
RHS Trade Marks are trademarks of the RHS and are being
used by the Publishers under licence from the Proprietor.

The Royal Horticultural Society is the UK's leading gardening charity
dedicated to advancing horticulture and promoting good gardening.
Its charitable work includes providing expert advice and information,
training the next generation of gardeners, creating hands-on
opportunities for children to grow plants, and conducting research
into plants, pests and environmental issues affecting gardeners.
For more information visit www.rhs.org.uk or call 020 3176 5800.

Managing Editor: Victoria Brooker
Design Manager: Peter Scoulding
RHS Books Publisher: Helen Griffin
RHS Head of Editorial: Tom Howard
RHS Book Editor: Simon Maughan
RHS Consultants: Mike Grant

ISBN: 978 1 5263 2853 3 (hbk)
ISBN: 978 1 5263 2854 0 (pbk)

Printed and bound in Dubai

MIX
Paper | Supporting
responsible forestry
FSC® C104740

Wayland,
an imprint of
Hachette Children's Group
Part of Hodder and Stoughton
Carmelite House
50 Victoria Embankment
London EC4Y 0DZ

An Hachette UK Company
www.hachette.co.uk
www.hachettechildrens.co.uk

PLANT FUN

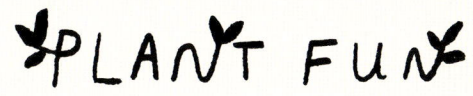

SUPER SEEDS

Susie Williams & Hannah Tolson

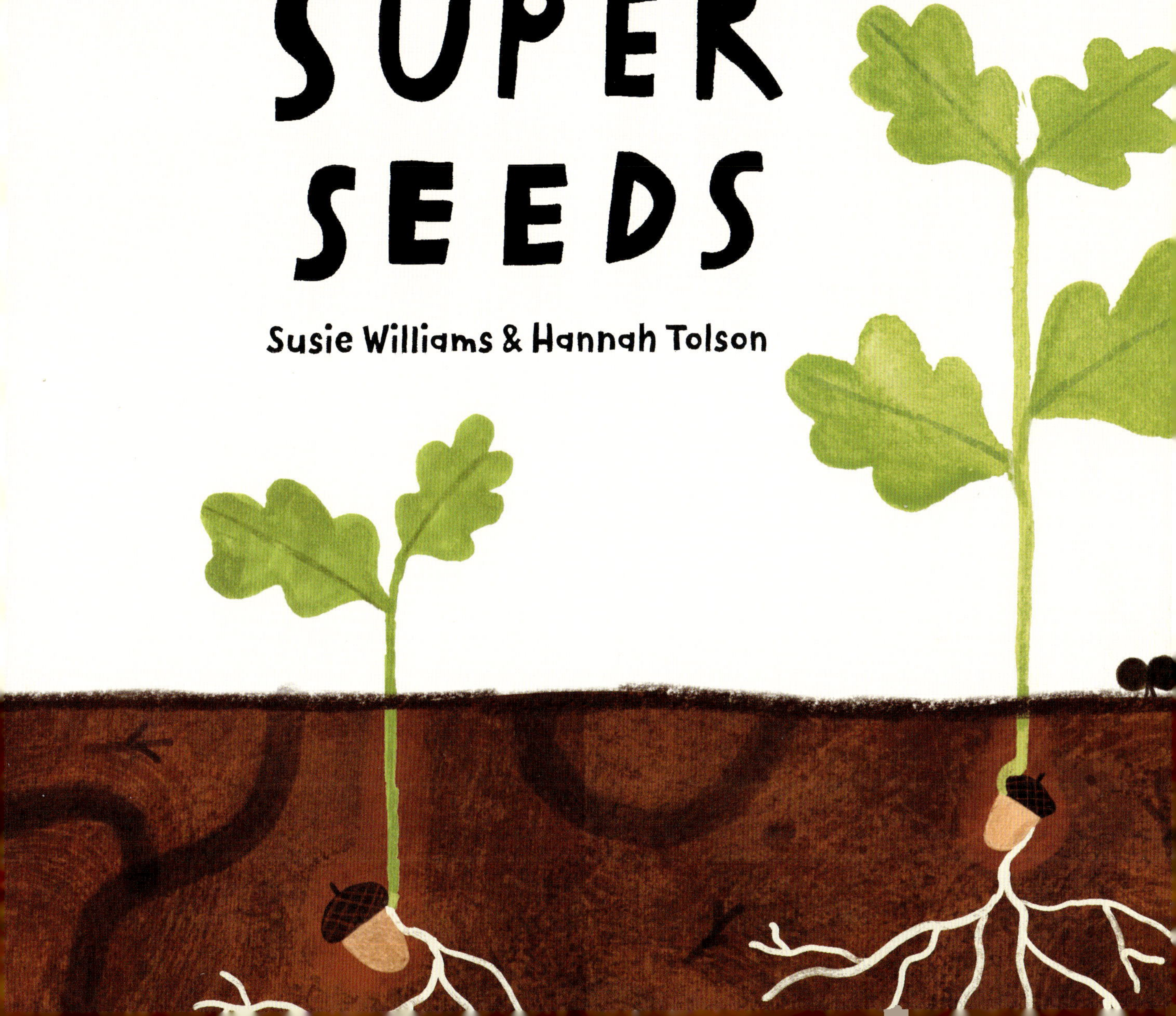

Plants grow in every part of the world.

From frozen lands to hot deserts, plants find ways to survive.

4

Trees, bushes, vegetables, ferns and mosses are all plants.

Most plants begin with a seed.

Seeds can be very small, such as the size of this full stop.

Some giant eucalyptus trees can grow from seeds this small.

The biggest seed is from the double coconut, which can be 50 cm wide.

seed

These enormous seeds grow on tall palm trees.

A seed contains everything
that a plant needs to grow.

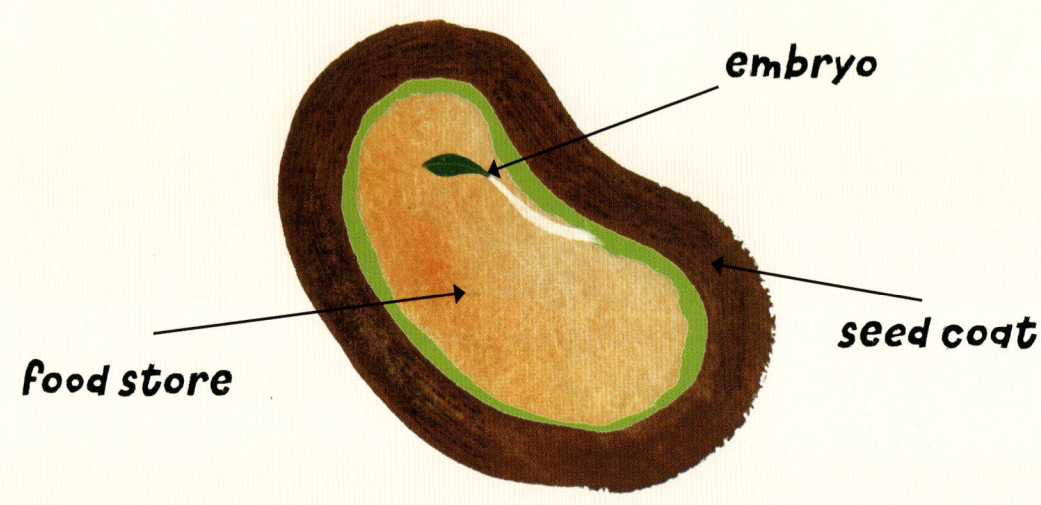

embryo

food store

seed coat

It contains the young
plant and some food.

For seeds to start growing, they need soil, water and the right temperature.

Seeds are how a plant
makes more plants.

When a seed is
planted, a root
grows down
into the soil.

The root holds
the seed
in place.

The tiny new plant, called a shoot, grows from the seed. The shoot pushes up out of the soil. When a seed starts to grow it is called germination.

The shoot grows tall
and strong until
it eventually flowers.

The flowers will turn into seeds.

Some flowers don't look like
the flowers in a garden.

Cones are actually the
flowers of some trees.
The seeds are between the
scales of the cone.

In grasses, the seeds are at
the top of the plant.

We eat the seeds from some grasses, such as wheat, oats and rice.

CEREAL

Some seeds are
inside fruits.

The seeds inside can all grow into new plants.

Some plants, such as the avocado plant, only have one seed in each fruit.

Pea pods might have three or four pea seeds in them.

One orchid's seed pod can contain 3 million seeds!

When the pumpkin-sized seed pod of the sandbox tree bursts, it hurls its seeds up to one hundred metres across the forest. The explosion echoes loudly through the undergrowth.

Some seeds, such as conkers, fall
to the ground and might start
growing there.

20

Others fly, such as seeds
from a sycamore tree.
The seeds are shaped like
helicopter wings to help
them travel further away.

If the seeds can travel, they
can find more space to grow.
This gives them a better
chance of becoming
a strong plant.

Seeds can travel in many different ways. Some animals eat the seeds that they find on the floor or on trees.

Squirrels bury acorns which might begin to grow into new oak trees.

Birds and deer eat the berries on bushes. Later, the seeds come out of the animals in their poo. The seeds can begin to grow.

The furthest that we know a seed
has travelled is about 4,800 km.

Coconuts float across seas
landing on islands far away.

The seeds are protected by the hard coconut shell. The shell has air in it to help it stay afloat.

Some seeds have developed hooks to help them cling on to animals that pass by.

As an animal brushes past a burdock plant, hooked seeds called burrs stick to the animal's fur.

The burr travels to
to a new place and falls off the
animal onto the ground.

When the time is right, the seed
will start growing into a new
plant. Then, one day, there will
be new seeds, too.

Germinate your own seeds

Seeds are amazing. Have fun planting your own seeds in a jar and watch as they start to grow.

You will need:
- a glass jar with a lid
- paper towels
- water
- 6-8 seeds, such as dried pinto beans or broad beans or peas

1. Dampen some paper towels with water. Make them wet but not soggy.
2. Place the paper towels in a jar.
3. Put the beans on the paper towels in the jar.
4. Screw the lid back on the jar.
5. Wait for the beans to start to germinate.

Watch as the seeds sprout shoots.
When the seeds have long sprouts much
bigger than the seeds, they are ready
to be planted in some soil.

Then you can watch your
new plant as it grows.

More wonderful facts about seeds

We can eat seeds, such as poppy, sunflower and pumpkin seeds. Beans, lentils, and nuts are also commonly eaten seeds. BUT *not all seeds are safe to eat.* Don't eat seeds that you find in the wild without checking with an adult.

Not all plants have seeds! Mosses and ferns produce spores instead of seeds.

In Wakehurst, UK, the Millennium Seed Bank has collected 2.4 billion seeds from different plants around the world to save them for the future in case plants become extinct.

Seeds can sense light and darkness!

Glossary

desert - a dry, sandy place

embryo - the beginning of an animal or plant

evergreen - a plant such as a tree or a bush that has green leaves on it all through the year.

sandbox tree - often found in the Amazon rainforest, the tree is also known as the dynamite tree because of its exploding seeds

seed coat - the outer layer around a seed

seed pod - the case that holds seeds

temperature - how hot or cold something is

Further information

You can visit the five Royal Horticultural Society gardens in person for all-year-round events, workshops, weekend and holiday fun! Find out about how to garden, discover the RHS Campaign for School Gardening, and visit the gardens virtually at www.rhs.org.uk/gardens

Discover the amazing plant collection at Royal Botanic Gardens, Kew ... www.kew.org/kew-gardens/plants ... and at the Millennium Seed Bank. www.kew.org/wakehurst/whats-at-wakehurst/millennium-seed-bank

Learn about the importance of plants with BBC Bitesize. www.bbc.co.uk/bitesize/topics/zxfrwmn/articles/zss9msg

Index

burrs 26, 27

cereal 15
coconut 24, 25
cones 13

desert 4
double coconut 7

ferns 5
flowers 12, 13
fruits 16, 17

germination 11, 28, 29
grasses 14, 15

mosses 5

roots 10, 11, 12

shoots 11
soil 9, 10, 11

trees 5, 22
  evergreen 13
  eucalyptus 6
  palm 7
  sandbox 19
  sycamore 21

temperature 9